EVALUATION OF SAFETY AND ENVIRONMENTAL METRICS FOR POTENTIAL APPLICATION AT CHEMICAL AGENT DISPOSAL FACILITIES

Committee on Evaluation of the Safety and Environmental Metrics for Potential Application at Chemical Agent Disposal Facilities

Board on Army Science and Technology

Division on Engineering and Physical Sciences

NATIONAL RESEARCH COUNCIL
OF THE NATIONAL ACADEMIES

THE NATIONAL ACADEMIES PRESS
Washington, D.C.
www.nap.edu

THE NATIONAL ACADEMIES PRESS **500 Fifth Street, N.W.** **Washington, DC 20001**

NOTICE: The project that is the subject of this report was approved by the Governing Board of the National Research Council, whose members are drawn from the councils of the National Academy of Sciences, the National Academy of Engineering, and the Institute of Medicine. The members of the committee responsible for the report were chosen for their special competences and with regard for appropriate balance.

This study was supported by Contract/Grant No. W911NF-08-C-0053 between the National Academy of Sciences and the U.S. Army. Any opinions, findings, conclusions, or recommendations expressed in this publication are those of the author(s) and do not necessarily reflect the views of the organizations or agencies that provided support for the project.

International Standard Book Number-13: 978-0-309-13092-9
International Standard Book Number-10: 0-309-13092-1

Limited copies of this report are available from

Board on Army Science and Technology
National Research Council
500 Fifth Street, N.W., Room 940
Washington, DC 20001
(202) 334-3118

Additional copies are available from

The National Academies Press
500 Fifth Street, N.W.
Lockbox 285
Washington, DC 20055
(800) 624-6242 or (202) 334-3313
(in the Washington metropolitan area)
Internet, http://www.nap.edu

Printed in the United States of America

THE NATIONAL ACADEMIES

Advisers to the Nation on Science, Engineering, and Medicine

The **National Academy of Sciences** is a private, nonprofit, self-perpetuating society of distinguished scholars engaged in scientific and engineering research, dedicated to the furtherance of science and technology and to their use for the general welfare. Upon the authority of the charter granted to it by the Congress in 1863, the Academy has a mandate that requires it to advise the federal government on scientific and technical matters. Dr. Ralph J. Cicerone is president of the National Academy of Sciences.

The **National Academy of Engineering** was established in 1964, under the charter of the National Academy of Sciences, as a parallel organization of outstanding engineers. It is autonomous in its administration and in the selection of its members, sharing with the National Academy of Sciences the responsibility for advising the federal government. The National Academy of Engineering also sponsors engineering programs aimed at meeting national needs, encourages education and research, and recognizes the superior achievements of engineers. Dr. Charles M. Vest is president of the National Academy of Engineering.

The **Institute of Medicine** was established in 1970 by the National Academy of Sciences to secure the services of eminent members of appropriate professions in the examination of policy matters pertaining to the health of the public. The Institute acts under the responsibility given to the National Academy of Sciences by its congressional charter to be an adviser to the federal government and, upon its own initiative, to identify issues of medical care, research, and education. Dr. Harvey V. Fineberg is president of the Institute of Medicine.

The **National Research Council** was organized by the National Academy of Sciences in 1916 to associate the broad community of science and technology with the Academy's purposes of furthering knowledge and advising the federal government. Functioning in accordance with general policies determined by the Academy, the Council has become the principal operating agency of both the National Academy of Sciences and the National Academy of Engineering in providing services to the government, the public, and the scientific and engineering communities. The Council is administered jointly by both Academies and the Institute of Medicine. Dr. Ralph J. Cicerone and Dr. Charles M. Vest are chair and vice chair, respectively, of the National Research Council.

www.national-academies.org

COMMITTEE ON EVALUATION OF THE SAFETY AND ENVIRONMENTAL METRICS FOR POTENTIAL APPLICATION AT CHEMICAL AGENT DISPOSAL FACILITIES

J. ROBERT GIBSON, *Chair*, Gibson Consulting, LLC, Wilmington, Delaware
RONALD M. BISHOP, AEHS, Inc., San Antonio, Texas
COLIN G. DRURY, State University of New York at Buffalo
JAMES H. JOHNSON, JR., Howard University, Washington, D.C.
RANDAL J. KELLER, Murray State University, Murray, Kentucky
W. MONROE KEYSERLING, University of Michigan, Ann Arbor
OTIS A. SHELTON, Praxair, Inc., Danbury, Connecticut
LEVI T. THOMPSON, JR., University of Michigan, Ann Arbor
LAWRENCE J. WASHINGTON, The Dow Chemical Company (retired), Midland, Michigan

Staff

MARGARET N. NOVACK, Study Director
JAMES C. MYSKA, Senior Research Associate
NIA D. JOHNSON, Senior Research Associate
ALICE V. WILLIAMS, Senior Program Assistant

BOARD ON ARMY SCIENCE AND TECHNOLOGY

Preface

By the end of 2009, more than 60 percent of the global chemical weapons stockpile declared by signatories to the Chemical Weapons Convention will have been destroyed, and of the 184 signatories, only three countries will possess chemical weapons—the United States, Russia, and Libya.

In the United States, destruction of the chemical weapons stockpile began in 1990, when Congress mandated that the Army and its contractors destroy the stockpile while ensuring maximum safety for workers, the public, and the environment. The destruction program has proceeded without serious exposure of any worker or member of the public to chemical agents, and risk to the public from a storage incident involving the aging stockpile has been reduced by more than 90 percent from what it was at the time destruction began on Johnston Island and in the continental United States.

While agent safety was of foremost concern during the initial years of destruction operations, the more traditional occupational safety and health programs were not emphasized as strongly as they should have been. The National Research Council's (NRC's) longtime Committee on Review and Evaluation of the Army Chemical Stockpile Disposal Program[1] (the Stockpile Committee) in a series of reports repeatedly encouraged the Army and its contractors to pay more attention to safety and to continuously improve its safety and environmental programs.

The Army and its contractors have responded to the Stockpile Committee's recommendations and have, commendably, improved safety performance at the chemical agent disposal facilities. At this time, safety at chemical agent disposal facilities is far better than the national average for all industries. Even so, the Army and its contractors are desirous of further improvement. To this end, the Chemical Materials Agency (CMA) asked the NRC to assist by reviewing CMA's existing safety and environmental metrics and making recommendations on which additional metrics might be developed to further improve its safety and environmental programs.

This report is the product of the NRC's response to the Army's request. As chair of the ad hoc Committee on Evaluation of the Safety and Environmental Metrics for Potential Application at Chemical Agent Disposal Facilities, I wish to thank my fellow committee members for their hard work and contributions to this report. The committee is grateful to the CMA for its scheduling of videoconferences and for its quick turnaround of committee questions to allow this report to be written in a timely manner. It is particularly grateful to Raj Malhotra, of the CMA, for facilitating the information gathering. The committee is also grateful to the Dow Chemical Company, Inc., and Praxair, Inc., for

[1] In 2006, the Committee on Review and Evaluation of the Army Chemical Stockpile Disposal Program was replaced with the current Committee on Chemical Stockpile Demilitarization.

their significant contributions to the committee's consideration of private sector safety and environmental metrics. Finally, the committee is grateful to the NRC staff for their assistance in gathering data, conducting research, and producing this report and shepherding it through the NRC report review process.

J. Robert Gibson, *Chair*
Committee on Evaluation of the Safety and Environmental Metrics for Potential Application at Chemical Agent Disposal Facilities

Acknowledgment of Reviewers

This report has been reviewed in draft form by individuals chosen for their diverse perspectives and technical expertise, in accordance with procedures approved by the National Research Council's (NRC's) Report Review Committee. The purpose of this independent review is to provide candid and critical comments that will assist the institution in making its published report as sound as possible and to ensure that the report meets institutional standards for objectivity, evidence, and responsiveness to the study charge. The review comments and draft manuscript remain confidential to protect the integrity of the deliberative process. We wish to thank the following individuals for their review of this report:

Joan B. Berkowitz, Farkas Berkowitz & Company,
F. Peter Boer, NAE, Tiger Scientific, Inc.,
Richard A. Conway, NAE, Union Carbide Corporation (retired),
Robert F. Herrick, NAE, Harvard School of Public Health,
Kenneth W. Kizer, IOM, Kizer & Associates, LLC, and
Jimmy L. Perkins, University of Texas Health Science Center.

Although the reviewers listed above have provided many constructive comments and suggestions, they were not asked to endorse the conclusions or recommendations nor did they see the final draft of the report before its release. The review of this report was overseen by Hyla Napadensky. Appointed by the NRC, she was responsible for making certain that an independent examination of this report was carried out in accordance with institutional procedures and that all review comments were carefully considered. Responsibility for the final content of this report rests entirely with the authoring committee and the institution.

Contents

APPENDIXES

Tables and Figure

TABLES

FIGURE

Acronyms and Abbreviations

AIChE	American Institute of Chemical Engineers
ANCDF	Anniston Chemical Agent Disposal Facility
AWFCO	automatic waste feed cutoff
BBP	behavior-based process
CDF	chemical agent disposal facility
CMA	Chemical Materials Agency (U.S. Army)
CO_2	carbon dioxide
DAWC	day away from work case
EBS	employee-based safety
EEA	environmental enforcement action
EH&S	employee health and safety
EPA	Environment Protection Agency
ESH	environmental, safety, and health
FAC	first aid case
GB	nerve agent (sarin)
H	mustard agent
HD	distilled mustard agent
HT	distilled mustard mixed with bis(2-chloroethylthioethyl) ether
JACADS	Johnston Atoll Chemical Agent Disposal System
LOPC	loss of primary containment
LWC	lost workday case
MACT	Maximum Achievable Control Technology
MVA	motor vehicle accident
NECDF	Newport Chemical Agent Disposal Facility
OSHA	Occupational Safety and Health Administration
PBCDF	Pine Bluff Chemical Agent Disposal Facility
RCI	root cause investigation
RCRA	Resource Conservation and Recovery Act
RI	recordable injury
RIR	recordable injury rate
RMTC	reportable medical treatment case
RWC	restricted work case
TOCDF	Tooele Chemical Agent Disposal Facility
UMCDF	Umatilla Chemical Agent Disposal Facility
VPP	Voluntary Protection Programs
VX	nerve agent

Summary

The U.S. Army's Chemical Materials Agency (CMA) is responsible for the destruction of the nation's chemical agent and munitions stockpile, except at two sites that fall under a separate Department of Defense program.[1] To meet this goal, CMA has built and operated incineration-based chemical agent disposal facilities (CDFs) on Johnston Atoll, in the Pacific Ocean; near Anniston, Alabama; Pine Bluff, Arkansas; Tooele, Utah; and Umatilla, Oregon. It has also built and operated neutralization-based CDFs near Aberdeen, Maryland, and Newport, Indiana. The CDFs on Johnston Atoll and Aberdeen have been closed, and the CDF near Newport is undergoing closure. The CDFs near Anniston, Pine Bluff, Tooele, and Umatilla are still in operation.

When Congress mandated the destruction of the chemical weapons stockpile, it specified that destruction operations must be executed with maximum protection for the workers, the public, and the environment. In the initial years of disposal operations, the National Research Council's longtime Committee on Review and Evaluation of the Army Chemical Stockpile Disposal Program, in its reports, repeatedly encouraged the Army and its contractors to pay attention to safety and to engage in continuous improvement.[2] The Army and its contractors have responded to this encouragement, and the operating CDFs enjoy exemplary safety records at this time. Table 2-1 gives site injury rates as of October 31, 2008, and Tables 2-2 and 2-3 provide environmental statistics for the sites. Even so, the Army and the CDFs are desirous of further improving safety and environmental performance and have asked the National Research Council to review the safety and environmental metrics used by the CDFs.

Specifically, the ad hoc Committee on Evaluation of the Safety and Environmental Metrics for Potential Application at Chemical Agent Disposal Facilities was established to carry out the following tasks:

- Review and evaluate existing safety and environmental metrics employed at CMA facilities,
- Examine commercial and industrial operations for potentially applicable safety and environmental metrics, and
- Assess new initiatives at national organizations (i.e., National Safety Council, Occupational Safety and Health Administration, etc.) that could be used by CMA.

As part of their ongoing effort to improve worker safety and environmental compliance, the CDFs employ a variety of metrics to measure performance and guide improvement efforts. Table 3-1 gives the categories of safety metrics used at the CDFs. The metrics include both leading indicators, which are forward looking and seek to identify problems before they occur, and lagging indicators, which are retrospective and lead

[1]The stockpiles at the Blue Grass Army Depot, in Kentucky, and the Pueblo Chemical Depot, in Colorado, fall under the Assembled Chemical Weapons Alternative Program.

[2]In 2006, the Committee on Review and Evaluation of the Army Chemical Stockpile Disposal Program was replaced with the current Committee on Chemical Stockpile Demilitarization.

to corrective action after injuries or incidents have occurred. This committee reviewed the status of safety and environmental programs and performance and the metrics used at the Anniston, Newport, Pine Bluff, Tooele, and Umatilla CDFs.

The committee noted that all CDFs engage in extensive data gathering, but the specific metrics derived from the data varied considerably from one facility to the other. This is not surprising, since each CDF has its own destruction mission, geography, and culture.

The committee gathered information on metrics and assessed new initiatives used by other government organizations, industry as a whole, and professional organizations, with an eye toward identifying metrics that might be useful to the CDFs. The government organizations surveyed include the Department of the Army (other than the CMA) and the Federal Aviation Administration. The private entities surveyed include the Center for Chemical Process Safety of the American Institute of Chemical Engineers (AIChE), Corning, Dow Chemical, Motorola, and Praxair. Many of the metrics employed by these organizations are detailed in Appendix B but not discussed in the body of the report. The committee believed that discussing the results of its fact finding and assessment would be more appropriate than recommending specific metrics.

The terminology used in this report is defined in the glossary that makes up Appendix A. While the definitions in the glossary may not necessarily conform to those of the CDFs or other organizations, the committee believes that they will afford the reader a clear idea of the meanings intended here.

For the reader's convenience, the committee's findings and recommendations, located in Chapter 5, are presented here as well.

Finding. Safety and environmental performance at the operating Chemical Materials Agency chemical agent disposal facilities has continuously improved and is currently significantly better than the national average industry as measured by lost workday cases and the recordable injury rate. Three of the five facilities are compliant with third-party accreditation requirements. All but one of the facilities have been certified with the Star designation by the Voluntary Protection Programs of OSHA and all conform to the International Organization for Standardization (ISO) 14001 environmental requirements.

Recommendation 1. The chemical agent disposal facilities should continue the process of continuous improvement to achieve levels of safety and environmental performance equivalent to those achieved by comparable industries. Third-party certifications should be continued and encouraged. All chemical agent disposal facilities should comply or obtain the Star designation of the OSHA Voluntary Protection Programs, and all should continue to comply with the most current ISO environmental management standards.

Finding. The terminology used to describe various metrics and related activities is not consistent across the chemical agent disposal facilities or within the Chemical Materials Agency. This makes it difficult to compare the sites in a meaningful way or to accurately analyze programwide data.

Recommendation 2. The Chemical Materials Agency should require the development of a system of clear and consistent definitions that can be applied across all chemical agent disposal facilities. Although each facility should have the flexibility to apply safety and environmental approaches that meet any unique needs, a particular metric should be defined consistently to allow for direct comparisons among the facilities.

Finding. The chemical agent disposal facilities collect extensive data on injuries, and most engage in some injury analysis. However, no facility takes full advantage of the data to create additional and potentially more sensitive metrics. The focus has been on lost workday cases and the recordable injury rate. Other possible metrics, such as medical treatment cases and first aid case rates, are not universally employed or communicated. The analyses simply list outcomes and incidental variables (e.g., department and day of week) and as such are not very useful metrics. Further, they fail to include some essential information such as the task being performed when the injury occurred and the location within the facility where it took place.

Finding. In addition to collecting data on injuries, all chemical agent disposal facilities collect extensive incident data, but there does not appear to be an incident investigation system that would enable the sites to analyze the data and extract from them indicators for preventive action. Insofar as they are reported, "metrics" are simple lists.

Finding. The chemical agent disposal facilities have many observation programs in place, but as with data on incidents, no metrics appear to be developed from them. Observations derived from the various programs are not combined or analyzed and, again, many of the reported metrics are simply lists.

Recommendation 3. The chemical agent disposal facilities should take full advantage of injury data to develop, employ, and communicate additional related metrics. All of the facilities should engage in injury analysis, and the analyses should include all relevant data and be structured so that meaningful indicators can be derived from them.

Recommendation 4. Incident data can be leading indicators for injuries, although they are also lagging indicators for conditions and behaviors that could result in injuries. The chemical agent disposal facilities should develop metrics from incident data—one such might be an unsafe acts index that could support the analysis of trends and point out a need for preventive action.

Recommendation 5. Chemical agent disposal facilities should stop reporting on and communicating data that are simple enumerations unless there is a clear understanding of the context for the data or a demonstrated connection to the continuous improvement of safety and/or environmental performance. For example, reporting absolute numbers of injuries by department conveys no information that can be translated into action, because the data have not been transformed into a metric that allows true department-to-department comparisons (i.e., departmental injury rates). Finally, the facilities should cease collecting data that are not used to develop metrics or meaningful indicators.

Finding. Key environmental metrics used by the chemical agent disposal facilities are based on the formal written notification that an applicable statutory or regulatory requirement promulgated by the Environmental Protection Agency or other authorized federal, state, interstate, regional, or local environmental regulatory agency has been violated. These metrics are lagging indicators.

Recommendation 6. The chemical agent disposal facilities should develop a broader set of leading environmental metrics. For example, incident reporting and analysis and observation programs could be extended to the environment area. Metrics could be developed that resemble leading safety metrics and could include the following:

- Projected use of energy, materials, and water;
- Time to correct violation and devise preventive action;
- Content of environmental training courses and frequency with which they are offered; and
- Observations of small spills or improper disposal of chemicals.

In addition, it is recommended that all available data be examined for patterns that might turn out to be useful leading indicators.

Finding. Metrics used at the chemical agent disposal facilities are mainly lagging ones that record relatively rare, undesirable outcomes such as recordable injuries. This practice does not yield good information on the real-time status of important leading variables such as physical conditions and work practices. As a result, workers and managers do not receive timely feedback on how well they are doing in maintaining a work environment that is free of conditions or behaviors that increase the risk of injury. Chapter 4 of this report provides examples of outstanding safety programs in the private sector and government. These programs focus on positive—that is to say, desirable—working conditions and practices, leading indicator variables, and the ongoing measurement of positive process variables.

Recommendation 7. Chemical agent disposal facilities should establish metrics that directly measure safety program effectiveness in near real time. These initiatives to establish metrics should focus on identifying leading variables that (1) set high standards for safe working conditions and (2) are a sign of a positive safety culture—for instance, 100 percent compliance in wearing personnel protective equipment; 100 percent compliance with correct use of lockout and tag out procedures; and the documented participation of management in the safety and environmental programs.

Recommendation 8. The chemical agent disposal facilities should conduct their own review of the best practices of the entities discussed in Chapter 4 to determine whether there are practices and metrics that would complement their own metrics and, in turn, benefit their own safety and environmental programs.

Finding. Chemical agent disposal facility processing operations generally consist of routine, repetitive, and much-practiced procedures. Safety will continue to be a key consideration as site activities transition to decommissioning, demolition, and handling and shipping of secondary wastes. Closure operations involve new and much more varied procedures. The award fee criteria may have different targets for closure because the current metrics and targets may not be appropriate for the closure phase.

Recommendation 9. The Chemical Materials Agency should establish a framework for developing metrics for the decommissioning and demolition of chemical agent disposal facilities. This framework should be used for all the facilities but on a site-specific basis. The framework should include safety and environmental metrics and targets, as well as a plan for communicating information to workers and the public. The metrics in use for operational processes should be reviewed for appropriateness and target levels. Additional metrics should be identified from the best practices for decommissioning and decontaminating industrial facilities and for the Environmental Protection Agency's Superfund program.

The following findings and recommendations might be useful for the CDFs to consider. The committee does not wish to prescribe these for all the CDFs because the degree to which they would be useful will vary based on each facility's safety and environmental culture, regulatory environment, and stage of agent processing. The committee believes that the management of each CDF can best weigh the potential utility of these recommendations.

Finding. Incidents are not classified and no metrics are derived from incident data.

Recommendation 10. Chemical agent disposal facilities should consider classifying incidents (one such class might be "incident with serious potential") to enable the development of additional metrics and help with prioritizing incident investigations.

Finding. None of the chemical agent disposal facilities develop or employ process safety metrics.

Recommendation 11. The chemical agent disposal facilities should consider developing and implementing leading and lagging metrics for process safety. They should consider using the American Institute of Chemical Engineers' Center for Chemical Process Safety document entitled *Process Safety Leading and Lagging Metrics* to guide implementation of process safety metrics.

1

Introduction

CHEMICAL WEAPONS STOCKPILE

For more than 50 years the United States has maintained an extensive stockpile of chemical weapons stored primarily in military depots across in the continental United States. Largely manufactured 40 or more years ago, the chemical agents and associated weapons in this stockpile are now obsolete.

The stockpile contains two types of chemical agents: cholinesterase-inhibiting nerve agents (GB and VX) and blister agents, primarily mustard (H, HD, and HT) but also a small amount of lewisite. These chemical agents, which are liquids at room temperature and normal pressures, are frequently and erroneously referred to as gases. Also included in the chemical weapons stockpile are bulk ("ton") containers and munitions. Types of munitions include rockets, mines, bombs, projectiles, and spray tanks. Many munitions contain chemical agent and energetic materials (propellants and/or explosives), a combination whose safe and efficient destruction poses particular challenges. Information on the location, size, and composition of the original continental U.S. stockpile is presented in Figure 1-1.

CHEMICAL WEAPONS DISPOSAL PROGRAM

The disposal of the chemical weapons stockpile is a major undertaking. In 1990, the stockpile included approximately 30,000 tons of chemical agents stored at eight chemical weapons depots operated by the Army in the continental United States.

In 1985, under a congressional mandate (Public Law 99-145), the Army instituted a sustained program to destroy elements of the chemical weapons stockpile. In 1992, when Congress enacted Public Law 102-484, the Army extended this program to destroy the entire stockpile.

Chemical weapons stored overseas were collected at Johnston Island, southwest of Hawaii, and destroyed at the Johnston Atoll Chemical Agent Disposal System (JACADS), the first operational chemical agent disposal facility (CDF). JACADS began destruction activities in 1990 and completed processing of the 2,031 tons of chemical agent and the associated 412,732 munitions and containers in the overseas stockpile in November 2000 (U.S. Army, 2001).

The largest share of the original continental U.S. stockpile (13,616 tons of agent) has been stored at the Deseret Chemical Depot near Tooele, Utah. This component of the stockpile is being processed by the Tooele Chemical Agent Disposal Facility.

Other disposal facilities—at Aberdeen, Maryland; Anniston, Alabama; Pine Bluff, Arkansas; Newport, Indiana; and Umatilla, Oregon—began destruction operations later and have collectively destroyed more than 55 percent of the original stockpile. At two sites, the Blue Grass Army Depot near Richmond, Kentucky, and the Pueblo Chemical Depot near Pueblo, Colorado, facility construction has only just begun and destruction operations have not yet started. Like JACADS, CDFs at two more sites—Aberdeen, Maryland, and Newport, Indiana—have completed their missions. The JACADS

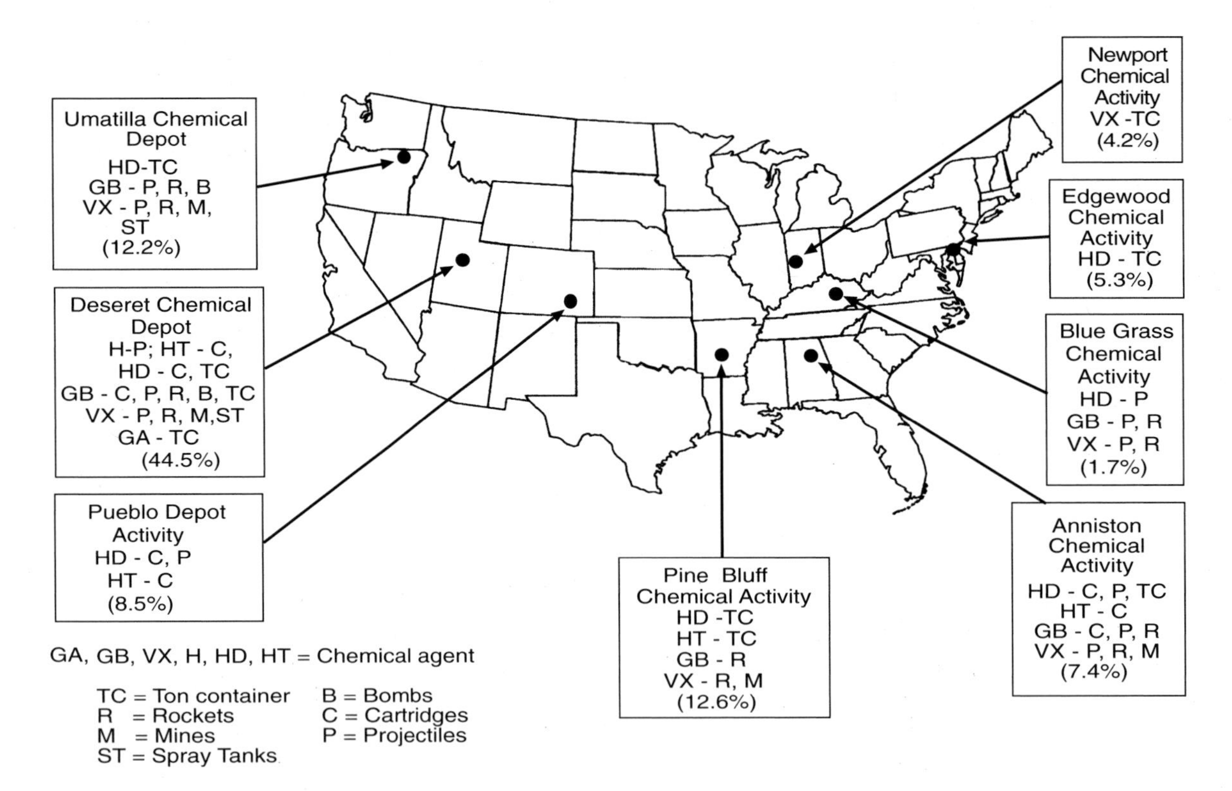

FIGURE 1-1 Location, size (percentage of the original stockpile), and composition of the eight continental U.S. storage sites. SOURCE: OTA, 1992.

and Aberdeen facilities have completed closure, and the Newport facility is entering closure.

SAFETY CHALLENGE

When Congress mandated the destruction of the chemical weapons stockpile, it specified that destruction operations must afford maximum protection to the workers, the public, and the environment. In the initial years of disposal operations the National Research Council's Committee on Review and Evaluation of the Army Chemical Stockpile Disposal Program, in its reports, repeatedly encouraged the Army and its contractors to pay attention to safety and to engage in continuous improvement.[1]

The Army and its contractors have responded to this encouragement, and, at the time of this writing, the five CDFs that are still operating have achieved recordable injury rates of about 1, compared with about 0.5 for the best industrial companies. The Army has expressed a desire and an intent to attain safety performance that equals or surpasses the performance of the best industrial companies. To help it reach this goal, the Army asked the National Research Council to review and evaluate the safety and environmental metrics employed at the operating facilities and, if necessary, to recommend additional metrics and/or program modifications.

STATEMENT OF TASK

The National Research Council will establish an ad hoc committee to:

- review and evaluate existing safety and environmental metrics employed at CMA facilities,
- examine commercial and industrial operations for potentially applicable safety and environmental metrics, and
- assess new initiatives at national organizations (i.e., National Safety Council, Occupational Safety and Health Administration, etc.) that could be used by CMA.

[1]In 2006, the Committee on Review and Evaluation of the Army Chemical Stockpile Disposal Program was replaced with the current Committee on Chemical Stockpile Demilitarization.

THE COMMITTEE, REPORT SCOPE, AND PROCESS

As is apparent from the statement of task, the committee was limited to considering safety and environmental metrics at the currently operating CDFs and the one that has entered closure. It does not address safety and environmental metrics at chemical stockpile storage sites, nor does it directly address population safety. Accordingly, a committee with very specific expertise was nominated to undertake this task (see Appendix D for biographical sketches of the committee members).

Two meetings were approved for this study. Consequently, the individual committee members needed to expend considerable effort between meetings. The first meeting was devoted to gathering information. The operating CDFs reported to the committee on the metrics they employ in managing their respective safety and environmental compliance programs and on the current status of these programs. Also, Chemical Materials Agency personnel provided a programmatic perspective on safety and environmental performance. A substantial portion of the reporting at the first meeting was done by videoconference. The situation at each facility as well as agency-wide is presented in Chapter 2.

After it had gathered the information for each facility, the committee analyzed the metrics and the manner in which they are used. The second meeting of the committee focused on this analysis and considered other metrics that might be employed by the various facilities. Committee deliberations were completed by means of a virtual meeting, wherein members talked via a teleconference and were able to view the report in real time on their computer screens. Chapters 3 and 4 present the committee's analysis, and Chapter 5 presents its findings and recommendations. The committee completed data gathering on October 31, 2008. A glossary of selected terms used in this report can be found in Appendix A. This glossary is intended only to clarify the meaning of a number of the terms in this report. It is not intended to create any definitions for adoption by the Chemical Materials Agency or the CDFs. The reader can also find a list of the acronyms used in this report immediately following the listing of tables and figures in the front matter of this report.

REFERENCES

OTA (Office of Technology Assessment). 1992. Disposal of Chemical Weapons: An Analysis of Alternatives to Incineration. Washington, D.C.: U.S. Government Printing Office.

U.S. Army. 2001. Status of Agent Destruction at JACADS and TOCDF, 5 September. Aberdeen Proving Ground, Md.: Program Manager for Chemical Demilitarization.

2

Summary of Current Safety and Environmental Metrics at Chemical Agent Disposal Facilities[1]

In this chapter, the collection, analysis, communication, and use of safety and environmental metrics are summarized, first at the level of the Chemical Materials Agency (CMA) and then at the level of the chemical agent disposal facilities (CDFs), in alphabetical order. These metrics are reviewed and evaluated in more detail in Chapter 3.

The term "safety metric" refers to a standard of measurement used to maintain an accident- and injury-free workplace, while "environmental metric" refers to a standard of measurement for chemical and handling processes relating to public health and the environment.

Metrics are characterized by:

1. What kinds of data are being collected—for example, number of injuries and number of regulatory noncompliances;
2. How the data are converted to metrics. This is typically done by dividing by a measure of exposure—for example, number of injuries per 200,000 working hours or number of regulatory noncompliance occurrences per month;
3. How this information is aggregated over departments or over time—for example, injury rates by department or by month or average annual rates of regulatory noncompliance;
4. How the information is used to improve safety —for example, detailed analysis of recent injuries to find opportunities for improvement or understanding the root causes of regulatory noncompliance to modify change procedures; and
5. How the information is communicated to management, safety professionals, the workforce, contractors, the public, and other sites.

Metrics can also be characterized by whether they are based on events that have already occurred, usually called lagging indicators, or on measured precursors to events, usually known as leading indicators. While lagging indicators give information of direct concern to management, the workforce, and the public, they can only be used for improvement after the fact. In contrast, leading indicators point the way to possible improvements in safety and environmental performance.

CHEMICAL MATERIALS AGENCY

CMA is responsible for the safe storage and destruction of most of the nation's chemical weapons stockpile. It oversees the activities in the five CDFs that are covered in this report. The headquarters management team, as well as scientific, communications, and support staff are based at the Edgewood Area of the Aberdeen Proving Ground in Maryland. Each CDF is government owned and contractor operated.

While CMA is responsible for the destruction of the chemical weapons stockpile, its contractors are responsible for ensuring that the congressional mandate

[1]Information gathering for this report ceased on October 31, 2008. The most current information is available at http://www.cma.army.mil.

for safety is fulfilled. To encourage exemplary performance, CMA has established award fee criteria so its contractors can be financially rewarded for safety and environmental performance over and above contract values. These criteria establish five categories in which performance will be measured and establish percentage ranges for each. The performance categories are these:

- Management, 15-25 percent
- Safety and surety, 25-35 percent
- Environmental, 20-30 percent
- Cost performance, 10-20 percent
- Schedule performance, 10-20 percent

With 45 to 65 percent of the total available award fee based on safety and environmental performance, there is significant incentive for CMA's contractors to excel in these areas.

The award fee criteria identify specific safety and environmental performance criteria against which contractor performance is assessed. Two safety criteria are specified: cases with days away from work, also known as lost workday cases (LWCs), and recordable injury rates (RIRs). The specified criteria for environmental performance are regulatory compliance actions, notices of noncompliance, and required submittals. CMA has established scoring and rating systems to assess performance against the award fee criteria.

ANNISTON CHEMICAL AGENT DISPOSAL FACILITY[2,3,4]

The Anniston Chemical Agent Disposal Facility (ANCDF) is located near Anniston, Alabama. It began destruction operations in August 2003 and currently employs 761 people. Since commencing operations, 50 percent of the agent stockpile has been destroyed (1,127 tons of the nerve agents GB and VX, and mustard). The estimated completion date for chemical weapons destruction is 2012. This, however, does not include the disposal of hazardous and secondary waste.

Safety and Environmental Performance and Metrics

Safety statistics for ANCDF and the four other sites can be found in Table 2-1 and environmental statistics for them can be found in Tables 2-2 and 2-3. The safety and environmental performance record at ANCDF is excellent and, in the committee's opinion, is the result of a leading safety culture. Various metrics are used in an effort to achieve continuous improvement.

ANCDF uses a number of what it calls leading safety and health indicators:

- The number of assessments performed by safety professionals;
- The number of assessments performed by a safety representative, a first-line supervisor, or a member of an employee-led committee;
- The number of assessments involving review of job safety analysis;
- The number of supervisors attaining Safety Trained Supervisor Certification; and
- The number of open actions and near-miss reports.

The safety metrics reported by ANCDF as lagging indicators include these:

- The overall injury rate,
- The total RIR,
- The lost time injury rate,
- Hours without an LWC,
- Number of reportable cases,
- Number of LWCs,
- Number of inspections,
- Regulatory citations, and
- Near misses.

It should be noted that the 2008 overall injury rate was almost half that in 2007—4.84 versus 8.32. See Table 2.1 for safety statistics.

The environmental metrics reported by ANCDF include these:

- Surveillances,
- Self-reported noncompliances,
- Resource Conservation and Recovery Act (RCRA) remedial actions,

[2]Cheryl Maggio, Deputy Project Manager, Chemical Stockpile Elimination, "Chemical stockpile elimination project overview," presentation to the committee on September 24, 2008.

[3]Robert Brook, Safety Manager, URS, "Safety metrics presentation," presentation to the committee on September 24, 2008.

[4]Ralph Nolte, Environmental Compliance Manager, URS, and Brian Thrasher, Deputy Environmental Manager, URS, "ANCDF environmental metrics," presentation to the committee on September 25, 2008.

TABLE 2-1 Chemical Demilitarization Site Injury Rates as of October 31, 2008

Facility	Employee Hours Worked Since Last LWC (hr)	Current 12-Month RIR	Highest 1-Month RIR[a]	Lowest 1-Month RIR[b]	Highest 12-Month RIR[c]	Lowest 12-Month RIR[d]
ANCDF	3.9 million (850 days)	0.73	5.18	0.0	1.73	0.27
NECDF	1.2 million (392 days)	0.52	4.45	0.0	1.95	0.33
PBCDF	1.0 million (250 days)	1.05	3.32	0.0	1.15	0.63
TOCDF	5.7 million (1102 days)	1.28	14.54[e]/ 11.26	0.0	4.82	0.71
UMCDF	3.9 million (850 days)	1.16	3.83	0.0	2.13	0.95

NOTE: LWC, lost workday case; RIR, recordable injury rate.
[a]Worst 1-month RIR in entire facility operational history, as of October 31, 2008.
[b]Best 1-month RIR in entire facility operational history, as of October 31, 2008.
[c]Worst 12-month RIR in entire facility operational history, as of October 31, 2008.
[d]Best 12-month RIR in entire facility operational history, as of October 31, 2008.
[e]The higher number includes 11 cases of food poisoning that occurred at a safety celebration picnic. The lower number is calculated without these cases included.
SOURCE: Cheryl Maggio, Deputy Project Manager Chemical Stockpile Elimination, CMA, "Chemical stockpile elimination project overview," presentation to the committee on September 24, 2008; Personal communication between Raj Malhotra, Deputy, Mission Support Directorate, CMA, and Margaret Novack, NRC, study director, December 10, 2008.

TABLE 2-2 Number of Environmental Enforcement Actions over the Last Five Fiscal Years

Facility	Fiscal Year					Average	Standard Deviation	Total
	2004	2005	2006	2007	2008[a]			
ANCDF	1	0	1	5	3	2	2	10
NECDF	1	1	0	0	0	0.4	0.55	2
PBCDF	0	0	1	0	1	0.4	0.55	2
TOCDF	1	2	2	1	1	1.4	0.55	7
UMCDF	1	4	2	4	0	2.2	1.79	11
Average	0.8	1.4	1.2	2.0	1.0			
Standard deviation	0.45	1.67	0.84	2.35	1.20			
Maximum	1	4	2	5	3			

[a]As of October 31, 2008.
SOURCE: Drew Lyle, Chief, Environmental Office, CMA, "Environmental performance measurement," presentation to the committee on September 24, 2008.

- Automated waste feed cutoffs (AWFCOs) and engineering stop feeds,
- Nonregulatory inspections, and
- Maximum Achievable Control Technology (MACT) exceedences.

Communication of Metrics

Information on safety metrics, as well as other safety information, is made available to manage-

TABLE 2-3 Environmental Noncompliances by Site

Facility	Low	High	2008[a]
ANCDF	5	45	5
NECDF	2	11	1
PBCDF	5	16	8
TOCDF	1	2	11
UMCDF	16	52	13

[a]As of October 31, 2008.
SOURCE: Drew Lyle, Chief, Environmental Office, CMA, "Environmental performance measurement," presentation to the committee on September 24, 2008; information provided by CMA.

ment from Prostat.[5] Information reported at ANCDF includes the number of reportable cases, the number of LWCs, the hours since the last LWC, the number of inspections, regulatory citations, and near misses, and an explanation of any Occupational Safety and Health Administration (OSHA) recordable injury (RI). *The Safety Digest* is a newsletter sent to the supervisory team so that members can disseminate safety information at employee meetings. Monthly injury statistics are used internally by the safety department, while statistical process control data on injury trending are made available to senior ANCDF management. Additionally, self-evaluation results under the Voluntary Protection Programs (VPP) are supplied to the regional VPP administrator as a measure of the safety program's effectiveness in meeting the stringent requirements established by OSHA's VPP.

Environmental information is provided through plan-of-the-day reports and schedule analysis packages. Additional environmental information is supplied for meetings of the Team for Environmental Awareness Compliance and Health and the Non-Compliance Review and Validation Squad.

[5]Prostat is a statistical analysis and data presentation tool used at all four CMA incineration sites. See http://www.polysoftware.com/stat.htm for more information.

NEWPORT CHEMICAL AGENT DISPOSAL FACILITY[6,7,8]

The Newport Chemical Agent Disposal Facility (NECDF) is located in Newport, Indiana. It was constructed between November 2000 and July 2003 and began destruction operations in May 2005. It currently employs 454 people. During its operation, 100 percent of its agent stockpile (1,269 tons of VX stored in 1,690 ton containers) was destroyed by caustic hydrolysis. All agent and waste products generated by agent hydrolysis have been disposed of, and it is now in the closure phase.

Safety and Environmental Performance and Metrics

The site safety statistics for NECDF can be found in Table 2-1 and the environmental statistics in Tables 2-2 and 2-3. During the operational phase of the facility (through August 2008), a total of 25,900 employee-based safety (EBS) observations were made. Three hundred and thirty-eight of the observations noted behaviors that could have placed the employees and people around them "at risk," and the remainder noted safe behaviors.

The safety and environmental metrics in place during closure could reasonably be expected to differ from the metrics during the operational phase of a facility. Although it is reasonable to assume that certain key metrics used during operations will continue to be employed during closure, no specific set of closure metrics was reported to the committee. The following discussion examines metrics that were used during the operational phase.

On a daily basis, the safety metrics included lost workdays, RIs, first aid cases (FACs), and days since last FAC. The safety metrics reported every week included lost workdays, RIs, FACs, days between FACs, near misses, injury by location on the body, number of EBS observations, number of findings of safe behavior, number of findings of "at risk" behav-

[6]Cheryl Maggio, Deputy Project Manager, Chemical Stockpile Elimination, "Chemical stockpile elimination project overview," presentation to the committee on September 24, 2008.

[7]Tulanda Brown, Risk Management Quality Assurance Director, Parsons, "NECDF safety metrics," presentation to the committee on September 24, 2008.

[8]Scott Rowden, Environmental Manager, Parsons, "Environmental metrics at NECDF," presentation to the committee on September 25, 2008.

ior, and safe behavior ratio. The monthly metrics include both lagging indicators (EBS observations) and leading indicators (supervisor safety inspections, safety contacts, and management observations). The facility manager said there were trigger points for action for certain metrics—for example, an RIR greater than 1.0.

During the facility's operational phase, three safety and environmental improvement programs were initiated. Ninety-five observers were trained for the EBS program. The management observation program was set up as a three-tier integrated system, and all supervisors and managers participated in human performance training.

Communication of Metrics

Information about safety and environmental metrics is communicated via a safety Web site, sitewide safety committees, and all-hands meetings. The meetings emphasize injuries and lessons learned and recognize and reward exemplary behavior and good safety metric performance.

PINE BLUFF CHEMICAL AGENT DISPOSAL FACILITY[9,10,11]

The Pine Bluff Chemical Agent Disposal Facility (PBCDF) is located at the Pine Bluff Arsenal in Pine Bluff, Arkansas. It began destruction operations in March 2005 and currently employs 773 people. Since commencing operations, 16.4 percent of the agent stockpile has been destroyed (631 tons out of the original total 3,850 tons of GB, VX, and mustard; PBCDF is preparing to dispose of mustard stored in ton containers). The estimated completion date for chemical weapons destruction is December 2011. This does not, however, include the disposal of hazardous and secondary waste.

[9]Cheryl Maggio, Deputy Project Manager, Chemical Stockpile Elimination, "Chemical stockpile elimination project overview," presentation to the committee on September 24, 2008.

[10]Marty Buell, Washington Demilitarization Company, Safety Manager, URS, "Safety metrics presentation," presentation to the committee on September 24, 2008.

[11]Greg Thomasson, Washington Demilitarization Company, Environmental Manager, URS, "PBCDF environmental metrics," presentation to the committee, September 25, 2008.

Safety and Environmental Performance and Metrics

PBCDF safety statistics can be found in Table 2-1 and environmental statistics in Tables 2-2 and 2-3. PBCDF employs a variety of safety and environmental metrics to continuously improve its safety and environmental programs. The safety metrics are compiled daily, weekly, monthly, quarterly, and annually. The daily metrics include a contract deliverable that provides a short description of any safety-related events that have occurred during the previous 24 hours. Weekly metrics include FACs, RIs, and near misses.

Monthly metrics include a contract deliverable that is distributed to management. It summarizes hours worked, cases reported, OSHA RIs, LWCs, and FACs. The facility tracks monthly injury trends by body part or cause. Leading indicators are also compiled monthly, including safety assessments that are performed by management, safety professionals, first-line supervisors, or by employee representatives.

Quarterly and annual tracking of metrics involves a compilation of injury trends, near misses, and safety observations that are submitted to management and the regional VPP administrator. All of the safety data collected are utilized to develop strategies for continuous improvement in safety performance at PBCDF.

The environmental metrics at PBCDF aim to minimize environmental enforcement actions and enhance the environmental culture at the facility. The facility reports site metrics weekly and reviews award fee metrics monthly with the project field office. These metrics include an assessment of the environmental culture and environmental compliance. The environmental culture is measured using a system that subjectively weighs environmental management system certification, training conducted, audits conducted, innovations, and articles published in an employee newsletter. Environmental compliance is measured using a system that weighs enforcement actions and both major and minor noncompliances. PBCDF examines other metrics as well, including self-reported noncompliances, environmental surveillances, and RCRA remedial actions. Facility staff also collects environmental data that are not transformed into metrics, including RCRA information (e.g., AWFCOs) and Clean Air Act information (e.g., fuel usage and furnace operating conditions). The facility's environmental management system conforms to the International Organization of Standardization (ISO) 14001 series of standards.

Communication of Metrics

Metrics are communicated at PBCDF in a number of ways. The facility publishes a weekly newsletter for employees that includes information on the safety record of the facility and any safety-related events. Quarterly and annual compilations of injury trends, near misses, and safety observations are communicated to management and the regional VPP administrator. All of the environmental metrics are communicated to an employee environmental leadership committee, the plant management, and the project manager.

TOOELE CHEMICAL AGENT DISPOSAL FACILITY[12,13,14]

The Tooele Chemical Agent Disposal Facility (TOCDF) is located at the Deseret Chemical Depot, near Tooele, Utah. It began destruction operations in August 1996 and currently employs 1,020 people. Since commencing operations, 71.8 percent of the agent stockpile has been destroyed (9,593 of 13,361 tons of GB, VX, and mustard; TOCDF is currently destroying mustard stored in ton containers). The estimated completion date for chemical weapons destruction operations is March 2012. This does not, however, include the disposal of hazardous and secondary waste.

Safety and Environmental Performance and Metrics

TOCDF safety statistics can be found in Table 2-1 and environmental statistics in Tables 2-2 and 2-3. TOCDF uses a variety of metrics to assess the performance of its safety and environmental programs. These metrics include both lagging indicators (e.g., RIs and environmental events) and leading indicators (e.g., near misses, observations, and inspections). These metrics are compiled and reported daily, weekly, monthly, quarterly, and annually.

Lagging safety metrics at TOCDF include counts of LWCs, RIs, and FACs. For recordable injuries, an RIR is calculated each month and used to update the 12-month rolling-average RIR for the facility. Days since the last RI and safe work hours (time since the last LWC) are updated and reported regularly to TOCDF personnel.

TOCDF also utilizes a number of leading metrics. An employee-based safety observation program has been implemented; the resulting metric is the number of observations performed per month, reported on a trend chart. The TOCDF Safety Department performs zone inspections to identify unsafe physical conditions and unsafe work practices. The number of zone inspections completed per month is used as a metric, and monthly histograms are produced showing counts of specific unsafe conditions and work practices. In addition to injury metrics, TOCDF also counts safety near misses and reports this metric on a weekly trend chart.

TOCDF performs total injury analysis on all RIs and FACs. Univariate histograms break down injury counts by a variety of categories (e.g., department, shift, injury type, and day of week). The facility also tracks the aging of safety work orders, including the number of open work orders less than 45 days old and the number of open work orders more than 45 days old. TOCDF submitted a VPP application to the OSHA Regional Office in Denver on September 30, 2008.

Lagging environmental metrics at TOCDF include state-identified noncompliances and self-reported (RCRA and Title V) noncompliances. These are counted on a monthly basis and the 12-month rolling average is updated monthly.

Leading environmental metrics include counts and timing (day of week) of RCRA inspections (performed by operations personnel) and regulatory inspections performed by the TOCDF Environment Department. Inspection findings are summarized in histograms by category (e.g., noncontainerized waste and container integrity) on a weekly basis, and trend charts are generated. In addition, TOCDF counts environmental near misses and reports this metric on a weekly trend chart.

TOCDF tracks the aging of RCRA work orders on a weekly basis, including the number of newly opened work orders and work orders closed. In addition, the trend in number of still-open work orders is given by week. MACT alarms are counted weekly for the metal parts furnace, and weekly counts of the reason for the alarms are charted in a histogram. AWFCOs are tracked for each furnace.

[12]Cheryl Maggio, Deputy Project Manager, Chemical Stockpile Elimination, "Chemical stockpile elimination project overview," presentation to the committee on September 24, 2008.

[13]Paul Anderson, Safety Manager, EG&G, "Safety metrics," presentation to the committee on September 24, 2008.

[14]Elizabeth Lowes, Deputy General Manager for Closure Integration, EG&G, "TOCDF environmental metrics," presentation to the committee on September 25, 2008.

TOCDF is a self-certified ISO 14001 facility. Nonregulatory environmental metrics tracked by TOCDF include the following:

- Annual natural gas usage,
- Water usage,
- Scrap metal recycled,
- Paper recycled, and
- Secondary waste processing/disposal.

Communications at Metrics

TOCDF staff prepares a daily status review and a daily progress report. There is also a weekly newsletter that includes key rates and daily counts. Days since the last RI are included in the daily safety management report that is distributed to all management and through the weekly Safety Action Team publication *SAT Newsletter*. Safe work hours are communicated very broadly on large signs around the site, the TOCDF intranet, in weekly and monthly management reports, and to all employees on the Safety Action Team. Safety metrics are reviewed on a quarterly basis. While not a metric, safety-related lessons learned are published and distributed to the demilitarization community.

Site-level metrics are communicated to all employees on a weekly, monthly, and quarterly basis and are available on the TOCDF intranet. Metrics are presented to the Site Environmental Leadership Committee, Departmental Corrective Action Review Boards, and the Site Corrective Action Review Boards.

UMATILLA CHEMICAL AGENT DISPOSAL FACILITY[15,16,17]

The Umatilla Chemical Agent Disposal Facility (UMCDF) is located at the Umatilla Chemical Depot in Hermiston, Oregon. It began destruction operations in September 2004 and currently employs 819 people. Since commencing operations, 35.5 percent of the agent stockpile has been destroyed (1,319 out of 3,719 tons of GB, VX, and mustard). The estimated completion date for chemical weapons destruction is July 2011. This does not, however, include the disposal of hazardous and secondary waste.

Safety and Environmental Performance and Metrics

UMCDF safety statistics can be found in Table 2-1 and environmental statistics in Tables 2-2 and 2-3. The facility uses a variety of safety metrics. On a daily basis the site reviews injuries and illnesses, near misses, property/vehicle damage, first aid visits, recordables, and days worked since LWC. On a weekly basis it looks at the 12-month RIR and the LWC rate, FACs, near misses, and the total recordable rate. On a monthly basis it reviews the OSHA 300 log, the project's total work rolling RIR, operations and maintenance rolling RIR, operations and maintenance subcontractor RIR, hours and days without a LWC, and FACs.[18]

Additionally, plant managers and the safety manager conduct and document weekly safety inspections of targeted work areas. Department safety professionals also conduct and document weekly safety assessments. The goal is for 80 percent or more of the assessments to result in no findings, and 100 percent of any findings to be resolved within one week. First-line supervisors and the shift safety representative conduct monthly inspections of work areas under their control.

UMCDF also uses a variety of environmental metrics. The metrics reviewed include self-reported noncompliances, surveillances, AWFCOs and engineering stop feeds, RCRA aging open items, regulatory and internal inspections, and MACT exceedences.

Safety metrics are reviewed daily, weekly, and monthly. The reviews are used for the annual award fee program and to identify areas of opportunity. Areas of opportunity identified are

- Better reporting of near misses;
- Behavior modification to help reduce unsafe acts;
- More attention to detail; and
- Increased identification and hazard control when the work involves fingers and hands, focusing on sharp objects and bodily motion.

[15]Cheryl Maggio, Deputy Project Manager, Chemical Stockpile Elimination, "Chemical stockpile elimination project overview," presentation to the committee on September 24, 2008.

[16]Emily Milliken, Safety Manager, URS, "Safety metrics," presentation to the committee on September 24, 2008.

[17]Jim Wenzel, Environmental Manager, UMCDF, "UMCDF environmental metrics," presentation to the committee on September 25, 2008.

[18]The OSHA 300 log is the document where recordable injuries are noted and documented. For more information, see http://www.osha.gov/pls/oshaweb/owadisp.show_document?p_table=STANDARDS&p_id=12805.

Communication of Metrics

Safety metrics are communicated in a variety of ways. One is a daily injury and illness report containing information on near misses, FACs, and recordables for the last 24 hours. A weekly operations analysis includes a review of the 12-month rolling RIR along with metrics on which the award fee is based: RIRs, FACs, and near misses. A weekly safety synopsis is delivered to project management, the field office, and the corporate office. In addition, there is a monthly report, which is a contract deliverable, including data from the OSHA 300 forms, a summary of near misses and FACs, and contract data requirements. A monthly corporate report is provided to the entire site via a Web site. A monthly report on injury trends by department is posted to the Web site and sent to management for use in identifying injury trends and controlling injuries. The annual trend report posted to the Web site contains information on root cause, hazard category, body part, day of week, time of injury, shift, department where near misses occur, and injuries and illnesses.

Environmental metrics are communicated to the Environmental Process Improvement Team, to quarterly meetings of supervisors and the project general manager, to "welcome back" briefings every Tuesday, and articles in *Today*. [19]

---

[19]*Today* is UMCDF's internal communication document.

3

Review and Evaluation of Metrics Currently Used at Chemical Agent Disposal Facilities

When representatives of the chemical agent disposal facilities (CDFs) and the Chemical Materials Agency (CMA) gave their presentations it became apparent that the terminology used to name and describe the metrics was not the same at all sites. Committee members were aware that they, too, might have even different definitions for a given metric. This diversity in definitions was also the case for much of the information reviewed by the committee. For example, the incidents, actions, and conditions that were categorized as near misses were very different at the different sites and did not conform with the CMA definition or with definitions found in reference materials. Thus, at one site, an unsafe act was categorized as a near miss, while at another it was categorized as an at-risk behavior but not a near miss. Because of this inconsistency in terminology, the committee developed a glossary (Appendix A) for use with this report. While the definitions found in the glossary may not agree with the definitions used by all, some, or even any of the CDFs or external organizations, the committee believes that they will afford the reader a clear and consistent basis for understanding the terminology used in this report.

SAFETY METRICS

Not surprisingly, each CDF has its own approach to safety and environmental programs, including the metrics it employs. All facilities, however, employ the two metrics that are specifically referenced in the Army's award fee criteria: the recordable injury rate (RIR) and lost workday cases (LWCs). The latter are cases with days away from work, as specified in the criteria document.

Beyond these two core injury metrics, each site accumulates data and develops metrics in accordance with its particular site safety program. The terminology employed at the various sites relates to

- Injuries (including illnesses),
- Incidents,
- Observations, and
- Miscellaneous metrics and activities.

In general, injuries and incidents are viewed as lagging indicators, observations are viewed as leading indicators, and miscellaneous metrics and activities can be viewed as one or the other. The types of metrics and activities encountered at the five CDFs are presented in Table 3-1.

The Occupational Safety and Health Administration (OSHA) requires that all injury data be captured at all sites. Even so, not all of the accumulated data are converted into metrics or used as such. All the injury metrics are lagging indicators and are useful for tracking performance and taking corrective action. Some, however, can be used to enhance and sustain awareness and could be viewed as leading indicators. For example, all of the CDFs use the injury metric "hours since the last LWC," whereas only the Tooele Chemical Agent Disposal Facility (TOCDF) utilizes the metric "time since the last injury"—specifically, days since last recordable injury (RI). The hours since the last LWC metric certainly enhances awareness and site pride with

TABLE 3-1 Types of Safety Metrics Employed at Chemical Agent Disposal Facilities

Injuries and Illnesses	Incidents	Observations	Miscellaneous Metrics and Activities
LWC	Near misses	Safety observations	Safety Trained Supervisor Certification[a]
LWC rate	At-risk behaviors	Employee observations	Corrective action and closure tracking
Hours since last LWC	Unsafe acts	Safety Department observations	Aborts of entries while wearing demilitarization protective ensemble
Restricted work case	Injury near misses[b]	Management observations	Lessons learned
Restricted work case rate		Supervisor safety observations	Safe behavior ratio
Medical treatment case		Stop-work orders	Total safe behaviors
Medical treatment case rate		Safety inspections	Compliance with OSHA Voluntary Protection Programs (VPP)
Total RIs		Safety assessments	Injury analysis
RI rate		Housekeeping	Incident analysis
Days since last RI		Unsafe conditions	
12-month rolling RIR		Substandard conditions	
First aid case (FAC)		Program audits	
FAC rate			
Days between FACs			
Days since last injury			

[a]At some facilities this is the number of supervisors who have attained this certification. At other facilities, it is the percentage of supervisors who have done so.

[b]This is an incident that nearly resulted in an injury, but did not.

respect to one type of injury, but it does not speak to other types of injuries, which in a true safety culture are equally important.

All of the CDFs engage in injury reporting, investigation, and analysis but not all develop metrics for preventive strategies based on the analyses. The data collected do not include much description of the facility location or the task being carried out at the time of injury, either or both of which could lead to identification of causal factors. Also, typical analyses rely on one-dimensional classifications of the outcome data based on conventional variables such as body part, injury type, and day of the week. There is no evidence of an active search for patterns in the data to find common causes that would lead directly to management action.

While all CDFs collect incident data as well as injury data, there does not appear to be an incident investigation system to gain the same depth of information as from injury investigations. Furthermore, it is unclear what triggered different depths of incident investigation. A classification system for incidents would help to ensure that precursors of injuries are better controlled and could lead to the development of additional metrics. Incidents can be leading indicators of injuries, although they are lagging indicators of the conditions and behaviors that can lead to injuries.

The safety observation programs in place at all CDFs vary in form and content. The five types of observation programs used across the facilities are

- Safety observations,
- Employee observations,
- Safety Department observations,
- Management observations, and
- Supervisor safety observations programs.

All facilities employed at least one type of program, but only one facility employed all five types. Interestingly, not one of the facilities said it had developed consolidated metrics from its multiple observation programs. Additionally there was no evidence that any CDF had validated its observation methodologies as a means of identifying precursors of incidents and injuries. All facilities characterized their observation programs as leading indicators, but this would be true only if the observation methodologies had been validated. Safety assessments, inspections, audits, and the like need to focus more on their findings than on the number of activities conducted.

Miscellaneous metrics are quite diverse, and not all have been included in Table 3-1. As was noted for metrics in the observations category, many of these are simple enumerations of actions or activities. For example, Safety Trained Supervisor Certification at one site is the number of individuals who have been trained in safety. At another site, the term refers to the percentage of individuals who have been so trained. While such metrics are very good measures to allow assessment of performance and/or compliance, they are not necessarily indicative of the actions a facility needs to take to improve the safety culture, remedy the physical conditions that create safety hazards, and/or modify behaviors that place personnel at increased risk of injury. Furthermore, a simple count of the number of supervisors who have been certified could be a misleading metric. Instead, metrics are needed that directly measure the extent to which a supervisor has established a strong safety culture and a safe work environment.

Overall, while lagging indicators dominate the safety metrics at all sites, they have been useful in advancing safety performance to its present state of excellence. Even so, sufficient data are accumulated to enable the development of additional metrics of both lagging and leading varieties to further improve the safety and environmental programs. This is especially true for leading indicators and/or metrics.

While the CDFs said they are using some leading indicators and/or metrics and are actively working to identify others, the committee believes that in general meaningful leading indicators are lacking. Most of the metrics that were proffered as leading indicators were little more than simple enumerations of actions and activities.

ENVIRONMENTAL METRICS

The CMA headquarters tracks several environmental metrics, including the number of environmental enforcement actions (EEAs), exceedences of chemical agent release limits, and stop-work orders. Most are lagging metrics that do not appear to characterize the violations, although a few are leading metrics. The facilities have developed their own sets of metrics that in some cases correspond to CMA headquarters' metrics.

The EEAs are defined as formal, written notification that an applicable statutory or regulatory requirement promulgated by the Environmental Protection Agency (EPA) or other authorized federal, state, interstate, regional, or local environmental regulatory agency has been violated. The incidence of EEAs was highest for the Umatilla facility, closely followed by EEAs at the Anniston facility. Again, there were no details on the character of the actions. No specific information was provided on the thresholds for chemical agent releases to be documented as environmental metrics although information is routinely gathered from EPA Toxic Release Inventories for the various facilities. Stop-work orders are emergency orders to cease or reduce activities for the purpose of protecting the environment. Only orders imposed by the EPA or state environmental agencies are counted as reportable metrics; decisions by Department of Defense or U.S. Army officials, site managers, and workers to stop work are not included in the metric, but reports are encouraged by CDF management.

Measures taken by CMA headquarters to promote environmental stewardship include quarterly environmental data calls where EEAs, regulatory inspections and permits, solid waste annual reports, and environmental performance audit systems are discussed.

While the facilities tracked different environmental metrics, most tracked Resource Conservation and Recovery Act remedial actions and self-reported occurrences of noncompliance. That having been said, the level of detail in the information provided to the committee seemed to vary. For example, TOCDF provided information about the number of actions and also defined targets for the key metrics, while other sites provided less information. Environmental metrics that were reported as being tracked include (typically at a specific facility) these:

- Furnace utilization (for efficiency assessment),
- Bulk and shredded paper for recycling (an EPA-inspired metric), and
- Remediation work orders.

In response to written questions, TOCDF reported tracking the following nonregulatory metrics: green purchasing, recycling of scrap metal and paper, secondary waste processing and disposal, fuel usage, and water consumption.[1]

The metrics are communicated periodically (in some cases weekly) to several communities, including the facility's general workforce and management, by means of review meetings, training sessions, and newsletters. The apparent goal of these communication efforts is to provide evidence of improvements.

[1]Personal communication between Trace Salmon, Deputy Site Project Manager, TOCDF, and Monroe Keyserling, committee member, October 14, 2008.

4

Assessment of Other Metrics Potentially Applicable to Chemical Agent Disposal Facilities

To accomplish the last two items on its statement of task, the committee assessed the development and use of leading and lagging indicators by various government and private organizations, including any new initiatives that the committee learned about in the course of its fact finding. Although it conducted this assessment the committee developed no related recommendations for implementation by the chemical agent disposal facilities (CDFs). Each such facility is in a unique situation with respect to site-specific geography, agent and munitions to be disposed of, demographics, culture, management, and regulatory climate. As a result, each site needs to be able to determine the metrics that are most appropriate for it. A top-down prescription for a standard set of metrics to be used at all CDFs would be less than helpful in the committee's view. Instead, the committee presents a general overview of the types of metrics that might provide general guidance to the CDFs for the continuing development of their safety and environmental programs and culture.

The committee assessed metrics that are currently used by the Department of the Army and the Federal Aviation Administration. They also assessed those of the Center for Chemical Process Safety of the American Institute of Chemical Engineers (AIChE), Corning, Dow Chemical, Motorola, and Praxair. The discussion of applicable metrics that follows is based on information compiled in Appendix B.

Data on environment, safety, and health (ESH) matters have historically been collected to provide management with quantifiable outcomes such as actual and hidden costs, lost time, and worker illness and injury. These data were often tied to regulatory compliance and/or worker compensation costs. The data used were almost exclusively lagging indicators—that is, they were collected after an incident, to determine strategies based on recorded outcomes to prevent future incidents. While lagging indicators serve useful purposes, they need to be supplemented with leading indicators to ensure continuous improvement of an ESH program. Also, lagging indicators are of minimal use for an organization such as the Chemical Materials Agency (CMA), whose mission is changing, moving from disposal of chemical agents to plant closure and disposal of hazardous and secondary waste.

Companies and organizations like the CMA that want to be the best in the field have recognized the limitations of lagging indicators and are seeking to use leading indicators to anticipate possible incidents within the ESH categories. While the literature is replete with information on leading indicators, their actual implementation to achieve continuous improvement is limited at best. Notwithstanding this, leading indicator models appear to have certain elements in common:

- Identifying hazards through risk assessment,
- Communication,
- Training,
- Documentation,
- Periodic review by top management,
- Follow-up on findings and corrective action,
- Analysis of near misses,
- Sharing of lessons learned,
- Worker involvement, and
- Audits and assessments.

While the definitions of lagging and leading indicators varied from one organization to the next, when the combined information from these organizations was considered some common themes for developing a system for using leading indicators became apparent. The definitions developed by this committee (see Appendix A) capture the intent of all of the organizations reviewed. The entire committee agrees that measurement is the precursor to control, and that the usual lagging indicators (e.g., those used by CMA and reviewed in Chapter 3) should not be neglected as they ultimately lead to improvement in safety performance outcomes.

Any measurements and derived indicators must be part of an overall system—an environmental policy or a commitment to continuous improvement—if they are to be effective in driving improvement. This system must possess a control philosophy (whether it is called risk management or safety management or is a business strategy such as Six Sigma), the commitment of top management, and specific goals for each indicator.

Input measures are the precursors to good safety performance. They include ensuring that safety is designed into all equipment and procedures, timely and effective training for all personnel, setting and meeting individual safety goals, and completion of tasks set (e.g., corrective actions, preventive actions, and permits issued). These inputs do not in themselves guarantee a safe organization, but they are a sign of how ready an organization is to achieve and improve safety.

Process measures are indicators that the organization and its workforce are performing their duties in a safe manner. Again, they do not guarantee safety, but without such indicators, reported levels of safety may reflect chance avoidance of rare events rather than safety levels achieved as a result of design and control. Typical process measures of safety include the number of near misses or incidents, behavior-based safety observations, rates of compliance with written procedures, participation in pretask hazard assessments, and audits or assessments of task performance and workplace factors. Note that assessments can be self-assessments, which provide useful training and involve the workforce, or independent assessments, which provide unbiased assurance of the state of processes. Ideally, all of these leading process measures should have been validated against outcome (lagging) indicators to ensure that they are indeed necessary conditions for safety performance, a step that is often neglected. Process measures include the actual physical processes of each task on a production line as well as completion of all preventive maintenance associated with the production line. The task processes are often evaluated by means of a job hazard analysis/process analysis as required under the Occupational Safety and Health Administration Voluntary Protection Programs (VPPs). While preventive maintenance may not in itself be an ESH direct indicator, the failure of a mechanical system may lead to an event that can produce an injury, exposure, or environmental insult. To this end, keeping to the preventive maintenance schedule and monitoring the completion of the tasks in that schedule are leading indicators of ESH.

Analysis of measured data is vital to ensuring evidenced-based control of safety. Analyses could entail the more thorough investigation of incidents and/or near misses to ensure that the causative factors have been identified; application of standard methodologies such as root-cause analysis; generation of indices in terms of rates rather than absolute numbers; and application of quality control techniques for visualizing trends. Such analyses provide a bridge between the raw data and management action, so that management has a clearer understanding of what needs to be changed and the potential effects of its actions on safety performance.

Accountability at both the organizational and individual levels is essential. Many organizations require the evaluation of support for and actual performance of ESH matters in employee and supervisor appraisals. In the organizations that excel in ESH, accountability includes penalties for specifically defined substandard performance.

Overall, much can be learned from the practices of industrial and government organizations about the derivation and use of leading indicators that could be applied to chemical demilitarization operations. Again, because of the unique circumstances in which each chemical agent disposal facility finds itself, the committee generalized its assessment to assist facilities in furthering their safety and environmental programs and cultures. The International Civil Aviation Organization summarizes overall safety management as follows:

> A safety management system . . . , as a minimum, identifies safety hazards, ensures that remedial action necessary to maintain an acceptable level of safety is implemented . . . , provides for continuous monitoring and regular assessment of the safety level achieved . . . , and aims to make continuous improvement to the overall level of safety.[1]

---

[1]Elwyn Jordan, Aviation Safety Inspector, Federal Aviation Administration, "Introduction to safety management systems (SMS)," presentation to the Federal Air Regulation 135 Seminar on April 19, 2007.

5

Findings and Recommendations

Finding. Safety and environmental performance at the operating Chemical Materials Agency chemical agent disposal facilities has continuously improved and is currently significantly better than the national average industry as measured by lost workday cases and the recordable injury rate. Three of the five facilities are compliant with third-party accreditation requirements. All but one of the facilities have been certified with the Star designation by the Voluntary Protection Programs of OSHA and all conform to the International Organization for Standardization (ISO) 14001 environmental requirements.

Recommendation 1. The chemical agent disposal facilities should continue the process of continuous improvement to achieve levels of safety and environmental performance equivalent to those achieved by comparable industries. Third-party certifications should be continued and encouraged. All chemical agent disposal facilities should comply or obtain the Star designation of the OSHA Voluntary Protection Programs, and all should continue to comply with the most current ISO environmental management standards.

Finding. The terminology used to describe various metrics and related activities is not consistent across the chemical agent disposal facilities or within the Chemical Materials Agency. This makes it difficult to compare the sites in a meaningful way or to accurately analyze programwide data.

Recommendation 2. The Chemical Materials Agency should require the development of a system of clear and consistent definitions that can be applied across all chemical agent disposal facilities. Although each facility should have the flexibility to apply safety and environmental approaches that meet any unique needs, a particular metric should be defined consistently to allow for direct comparisons among the facilities.

Finding. The chemical agent disposal facilities collect extensive data on injuries, and most engage in some injury analysis. However, no facility takes full advantage of the data to create additional and potentially more sensitive metrics. The focus has been on lost workday cases and the recordable injury rate. Other possible metrics, such as medical treatment cases and first aid case rates, are not universally employed or communicated. The analyses simply list outcomes and incidental variables (e.g., department and day of week) and as such are not very useful metrics. Further, they fail to include some essential information such as the task being performed when the injury occurred and the location within the facility where it took place.

Finding. In addition to collecting data on injuries, all chemical agent disposal facilities collect extensive incident data, but there does not appear to be an incident investigation system that would enable the sites to analyze the data and extract from them indicators for preventive action. Insofar as they are reported, "metrics" are simple lists.

Finding. The chemical agent disposal facilities have many observation programs in place, but as with data on incidents, no metrics appear to be developed from them. Observations derived from the various programs are not combined or analyzed and, again, many of the reported metrics are simply lists.

Recommendation 3. The chemical agent disposal facilities should take full advantage of injury data to develop, employ, and communicate additional related metrics. All of the facilities should engage in injury analysis, and the analyses should include all relevant data and be structured so that meaningful indicators can be derived from them.

Recommendation 4. Incident data can be leading indicators for injuries, although they are also lagging indicators for conditions and behaviors that could result in injuries. The chemical agent disposal facilities should develop metrics from incident data—one such might be an unsafe acts index that could support the analysis of trends and point out a need for preventive action.

Recommendation 5. Chemical agent disposal facilities should stop reporting on and communicating data that are simple enumerations unless there is a clear understanding of the context for the data or a demonstrated connection to the continuous improvement of safety and/or environmental performance. For example, reporting absolute numbers of injuries by department conveys no information that can be translated into action, because the data have not been transformed into a metric that allows true department-to-department comparisons (i.e., departmental injury rates). Finally, the facilities should cease collecting data that are not used to develop metrics or meaningful indicators.

Finding. Key environmental metrics used by the chemical agent disposal facilities are based on the formal written notification that an applicable statutory or regulatory requirement promulgated by the Environmental Protection Agency or other authorized federal, state, interstate, regional, or local environmental regulatory agency has been violated. These metrics are lagging indicators.

Recommendation 6. The chemical agent disposal facilities should develop a broader set of leading environmental metrics. For example, incident reporting and analysis and observation programs could be extended to the environment area. Metrics could be developed that resemble leading safety metrics and could include the following:

- Projected use of energy, materials, and water;
- Time to correct violation and devise preventive action;
- Content of environmental training courses and frequency with which they are offered; and
- Observations of small spills or improper disposal of chemicals.

In addition, it is recommended that all available data be examined for patterns that might turn out to be useful leading indicators.

Finding. Metrics used at the chemical agent disposal facilities are mainly lagging ones that record relatively rare, undesirable outcomes such as recordable injuries. This practice does not yield good information on the real-time status of important leading variables such as physical conditions and work practices. As a result, workers and managers do not receive timely feedback on how well they are doing in maintaining a work environment that is free of conditions or behaviors that increase the risk of injury. Chapter 4 of this report provides examples of outstanding safety programs in the private sector and government. These programs focus on positive—that is to say, desirable—working conditions and practices, leading indicator variables, and the ongoing measurement of positive process variables.

Recommendation 7. Chemical agent disposal facilities should establish metrics that directly measure safety program effectiveness in near real time. These initiatives to establish metrics should focus on identifying leading variables that (1) set high standards for safe working conditions and (2) are a sign of a positive safety culture—for instance, 100 percent compliance in wearing personnel protective equipment; 100 percent compliance with correct use of lockout and tag out procedures; and the documented participation of management in the safety and environmental programs.

Recommendation 8. The chemical agent disposal facilities should conduct their own review of the best practices of the entities discussed in Chapter 4 to determine whether there are practices and metrics that would complement their own metrics and, in turn, benefit their own safety and environmental programs.

Finding. Chemical agent disposal facility processing operations generally consist of routine, repetitive, and much-practiced procedures. Safety will continue to be a key consideration as site activities transition to decommissioning, demolition, and handling and shipping of secondary wastes. Closure operations involve new and much more varied procedures. The award fee criteria may have different targets for closure because the current metrics and targets may not be appropriate for the closure phase.

Recommendation 9. The Chemical Materials Agency should establish a framework for developing metrics for the decommissioning and demolition of chemical agent disposal facilities. This framework should be used for all the facilities but on a site-specific basis. The framework should include safety and environmental metrics and targets, as well as a plan for communicating information to workers and the public. The metrics in use for operational processes should be reviewed for appropriateness and target levels. Additional metrics should be identified from the best practices for decommissioning and decontaminating industrial facilities and for the Environmental Protection Agency's Superfund program.

The following findings and recommendations might be useful for the CDFs to consider. The committee does not wish to prescribe these for all the CDFs because the degree to which they would be useful will vary based on each facility's safety and environmental culture, regulatory environment, and stage of agent processing. The committee believes that the management of each CDF can best weigh the potential utility of these recommendations.

Finding. Incidents are not classified and no metrics are derived from incident data.

Recommendation 10. Chemical agent disposal facilities should consider classifying incidents (one such class might be "incident with serious potential") to enable the development of additional metrics and help with prioritizing incident investigations.

Finding. None of the chemical agent disposal facilities develop or employ process safety metrics.

Recommendation 11. The chemical agent disposal facilities should consider developing and implementing leading and lagging metrics for process safety. They should consider using the American Institute of Chemical Engineers' Center for Chemical Process Safety document entitled *Process Safety Leading and Lagging Metrics* to guide implementation of process safety metrics.

Appendixes

Appendix A

Glossary

This glossary has been developed solely for use with this report. It should not be construed as a recommendation by the committee of a common set of definitions for the terms included, which are presented in conceptually related groups rather than alphabetically.

injury	physical trauma to a body part that requires treatment in some form
recordable injury	injury that because of the kind of treatment it requires is reportable to the Occupational Safety and Health Administration and is thus recorded in the OSHA 300 log
recordable injury rate (RIR)	number of recordable injuries per 200,000 hours worked
rolling RIR	12-month moving average of the recordable injury rate
lost workday case (LWC)[1]	injury that is sufficiently severe to require the injured person to miss at least one full day of work, not including the day the injury occurred
restricted workday case (RWC)[1]	injury that is sufficiently severe to cause the injured person to be unable to fully perform all of his or her normal job functions on a day(s) other than the day the injury occurred
medical treatment case (MTC)[1]	injury that is sufficiently severe to require substantial treatment and/or prescription medication by a medical professional but does not prevent the injured individual from performing his or her normal job functions
first aid case (FAC)[1]	injury that is not sufficiently severe to require more than minimal treatment and/or nonprescription medication
event	actual occurrence or happening

[1]The rate for such an injury is the number of injuries per 200,000 hours worked.

incident — event that could have resulted in an injury or property damage or both but that did not

incident with serious potential — incident that could have caused extensive injuries (including loss of human life) or substantial property losses or both but that did not

unsafe act — action by an individual that increases the risk of injury to himself or herself and/or to others

unsafe act index — ratio between the number of unsafe acts observed and the number of individuals observed

unsafe condition — physical, mechanical, or other condition that increases the risk of injury to individuals who are in proximity to the condition

observation — action, condition, incident, or event that was noted and documented by an individual or individuals during the course of a safety or environmental assessment, inspection, audit, or other safety or environmental program, whether scheduled or not

metric — measurement or system of measurements used to analyze and improve performance

injury/incident analysis — process of organizing injury or incident statistics by shared factors such as body part, injury type, time of day, task being performed, and location, for the purpose of spotting trends in occurrences

leading indicator — prospective metric or set of metrics that can be used to develop strategies for prevention of injuries and incidents

lagging indicator — retrospective metric or set of metrics that can point to a need for corrective action

Appendix B

Safety and Environmental Metrics Employed by Private Companies Surveyed for This Report

Table B-1 compiles the safety and environmental metrics used by the private companies surveyed for this report. These metrics and those of some other government organizations are discussed in Chapter 4.

TABLE B-1 Safety and Environmental Metrics Employed by Private Companies Surveyed for This Report

Measure	Area	Definition	Type	Comments
Number of recordable injuries (RIs) or illnesses	Personal safety		Lagging	Per OSHA requirements
Number of lost workday cases (LWCs)	Personal safety		Lagging	Per OSHA requirements
Contractor injury or illness rate	Personal safety	Number of RIs per number of work hours × 200,000	Lagging	RMTC, RWC, and DAWC (all OSHA definitions)
Company injury or illness rate	Personal safety	Number of RIs per number of work hours × 200,000	Lagging	RMTC, RWC, and DAWC (all OSHA definitions)
Near miss	Personal/environmental/ transportation/process safety	Number of unsafe conditions or events that almost injured someone but didn't or almost spilled something but didn't	Leading	Can identify unsafe conditions, safety incidents that could have been more serious in different circumstances, etc.
Corrective and preventive actions	Personal/environmental/ transportation/process safety	Proportion of corrective and preventive actions closed on time to total number of action items	Leading	Percent of action items related to employee health and safety (EH&S) incidents that have been closed by the due date
Behavior-based process (BBP) observation	Personal/environmental/ transportation/process safety	Number of observations of behavior as part of a behavior-based safety program	Leading	Total number of observations made of a work group in a given time
Percent safe BBP observations	Personal/environmental/ transportation/ process safety	Number of safe behaviors/ total behaviors	Leading	The percentage of safe behaviors should be less than 100 percent since your program should be looking at behaviors that you want to change and at behaviors that you are getting much better at

Continued

TABLE B-1 Continued

Measure	Area	Definition	Type	Comments
BBP observation—analysis to drive behavior change	Personal safety	Number of analyses performed	Leading	Should analyze the antecedents and consequences of an unwanted behavior at least quarterly. Behavior might be improved by adjusting an antecedent
BBP observation—driving behavior change	Personal safety	Number of critical behaviors that reached habit strength	Leading	Try to drive at least one behavior to habit strength per year by adjusting the antecedents and consequences of that behavior
Procedure use	Personal/process safety	Number of critical procedures used/number of critical procedure required tasks performed	Leading	Can be daily, weekly, or monthly depending on the size of the organization. Tasks that require a critical procedure are defined by the facility
Quality of root cause investigation (RCI)	Personal/environmental/transportation/process safety	Number of minimum quality criteria met for the RCIs in a given period	Leading	RCI minimum criteria are defined by the company
Pretask hazard assessment participation	Personal safety	Number of pretask hazard assessments performed	Leading	Assessment can be conducted per person or per work group, weekly or monthly
Performance tracking on permits	Personal safety	Number of defects found per permit	Leading	Permit documentation is audited and any mistake or omission is a defect (safe work permit/isolation of energy/confined space entry)
Training timeliness	Personal/environmental/transportation/process safety	Required training completed on time—not overdue.	Leading	Overdue EH&S training is a sign of a slipping safety culture and priority.
Compliance task tool	Personal/environmental/transportation/process safety	Number of required compliance tasks overdue/total number of required compliance tasks	Leading	Overdue safety compliance tasks are a sign of slipping safety culture and priority. An example of these tasks is fire extinguisher inspections.
Severity rate	Personal safety	Number of (RMTC × 1) + (RWC × 3) + (DAWC × 9) + (fatalities × 27) per 200,000 work hours	Lagging	Gives a weighted rate.
DAWC count	Personal safety	Number of DAWCs	Lagging	
DAWC rate	Personal safety	Number of DAWC per 200,000 work hours	Lagging	
Loss of primary containment (LOPC) count	Personal safety	Number of LOPCs	Lagging	For example, leaks, breaks, and spills
Severe LOPC (Categories 1,1A, and 2A)	Personal/environmental/process safety	Number of Category 1, 1A, and 2A LOPCs	Lagging	Category 1 is any loss of primary containment resulting in the release of >5,000 lb flammable chemical. Category 1A is a release causing a DAWC. Category 2A is a spill resulting in a RI.
Category 4 LOPC count	Personal/environmental/process safety	Number of Category 4 LOPCs	Leading	Category 4 is a minor spill of <100 lb that has no measurable impact on people or the environment.
Ratio of Category 4 LOPC to Categories 1, 2, and 3 LOPCs	Personal safety	Ratio of Category 4 LOPCs to all other categories of LOPCs	Leading	Try to achieve a 40:1 ratio in order to find the small spills and fix them before they become larger spills. (Category 2 is a loss of primary containment with a release of >1,000 lb or an RMTC or a RWC (2A). Category 3 is any LOPC that loses >100 lb of chemical or 1,000 lb of dry inert solids).

TABLE B-1 Continued

Measure	Area	Definition	Type	Comments
Number of process safety events	Process safety	Number of events within a specified time period. The severity of events may be low, medium, or high.	Both lagging and leading	For near misses, it's a leading indicator.
Number of fatality potential events	Personal/transportation/ process safety	Number of such events within a specified time period	Lagging	Measure progress in addressing high potential events
Motor vehicle accident (MVA) count	Transportation	Number of MVAs	Lagging	An MVA is a motor vehicle accident resulting in personal injury or at least $500 in damage.
MVA rate	Transportation	Number of MVAs per million miles driven	Lagging	Includes all miles driven from company owned, leased, or rented vehicles and miles driven on company business from personal vehicles
Number of preventable accidents or number of preventable accidents per unit time or distance	Transportation	Number of preventable product-carrying vehicle accidents or a rate based on this number	Lagging	
Number of high-severity accidents or number of high-severity accidents per unit time or distance	Transportation	Number of high-severity product-carrying vehicle accidents or a rate based on this number	Lagging	
Number of rollovers/ rollover rate	Transportation	Number of product carrying vehicle rollovers or a rate based on this number	Lagging	
Energy intensity	Environmental	British thermal units per pound production	Lagging	
Greenhouse gas (GHG) energy efficiency	Environmental	Quantity of carbon dioxide (CO_2) generated per unit of production	Lagging	
Wastewater intensity	Environmental	Pounds of wastewater per pound of production	Lagging	Water that is treated at a wastewater treatment facility
Waste intensity	Environmental	Pounds of waste per pound of production	Lagging	Material that receives end-of-pipe treatment; report as the bulk amount prior to treatment.
Total waste weight	Environmental	Weight by type and disposal method	Lagging	
Chemical emissions	Environmental	Chemical emissions (tons)	Lagging	Material that is released to the environment that does not receive end-of-pipe treatment (not including water). Chemical emissions exclude conventional emissions such as combustion products (nitrogen oxides, carbon monoxide, sulfur oxides, CO_2, and particulates), methane, and hydrogen. Also excluded are the "normally excluded as an emission" compounds from GEI such as nitrogen, oxygen, water, aluminum, and salts (chlorides, sulfates, hydroxides, oxides, hypochlorite, and carbonates).

Continued

TABLE B-1 Continued

Measure	Area	Definition	Type	Comments
Priority compound emissions	Environmental	Priority compounds (tons)	Lagging	A list of priority chemicals that include persistent bioaccumulative and toxic compounds; selected known human carcinogens; selected ozone depletors; and high-volume toxic compounds
Volatile organic compound emissions	Environmental	Volatile organic compounds (tons)	Lagging	
Total water use	Environmental	Pounds or gallons water used/time period	Lagging	
Direct GHG emissions	Environmental	CO_2-equivalent metric tons	Lagging	Direct GHG emissions are those that are emitted from a company location. Direct emissions include all GHGs emitted from any on-site fugitive or air point source.
Kyoto GHGs as CO_2-equivalent intensity	Environmental	Pounds of CO_2-equivalent per pound production	Lagging	
Assessment compliance performance	Personal/environmental/transportation/process safety	Assigned grade to each area reviewed in assessment	Leading	Commonly understood measure for assessing improvement in performance
Percent of safety alerts completed	Personal/environmental/transportation/process safety	Percent completion by facilities covered by alerts	Leading	Drives implementation of lessons learned from safety incidents
Number of potential environmental noncompliances	Environmental	Internally reported potential environmental noncompliances per month	Leading	Proactive measure of effectiveness of environmental program
Number of significant environmental spills	Environmental	Spills per unit time	Lagging	
Toxic release inventory on site releases	Environmental	Number of releases per unit time	Lagging	

NOTE: RCI, root cause investigation; OSHA, Occupational Safety and Health Administration; RMTC, reportable medical treatment case; RWC, restricted work case; DAWC, days away from work case; LOPC, loss of primary containment; BBP, behavior-based process; RCI, root cause investigation; ES&H, employee safety and health; RI, recordable injury; GHG, greenhouse gas; GEI, greenhouse gas emissions.

SOURCE: Data provided by Corning, Dow Chemical, Motorola, and Praxair.

Appendix C

Committee Meetings

FIRST COMMITTEE MEETING, SEPTEMBER 24-26, 2008, WASHINGTON, D.C.

Objective: To receive detailed briefings on processes and equipment, review the preliminary report outline and the report writing process, confirm committee writing assignments, and decide future meeting dates and next steps.

Chemical Demilitarization 101, Cheryl Maggio, Deputy, PMCSE, Chemical Materials Agency

Chemical Materials Agency Safety Metrics, C.T. Anderson, Chief, Safety Office, Risk Management Directorate, Chemical Materials Agency

Safety Videoconference with safety staff at the Anniston, Newport, Pine Bluff, Tooele, and Umatilla CDFs.

Chemical Materials Agency Environmental Metrics, Drew Lyle, Chief, Environmental Office, Risk Management Directorate, Chemical Materials Agency

Environmental Videoconferences on environmental metrics with environmental staff of the Anniston, Newport, Pine Bluff, Tooele, and Umatilla CDFs

SECOND COMMITTEE MEETING, OCTOBER 28-30, 2008, WASHINGTON, D.C.

Objective: To develop text and refine the report. Only committee members and staff attended.

VIRTUAL COMMITTEE MEETING, JANUARY 6, 2009

Objective: To discuss the report draft, resolve remaining issues, and generate a document that is ready for concurrence. This meeting was conducted over the Web, with document editing carried out online and in real time and an accompanying teleconference.

Appendix D

Biographical Sketches of Committee Members

J. Robert Gibson, *Chair*, retired as a director of DuPont's Crop Protection Products Division. During his 30-year career with DuPont, Dr. Gibson held positions in R&D, chemical plant management, and corporate administration (at one point, he was Corporate Director of Safety and Health). He was also assistant director of DuPont's Haskell Laboratory for Toxicology and Industrial Medicine. He was certified in toxicology by the American Board of Toxicology from 1980 to 2005 and is currently a consultant in toxicology and occupational safety and health. Dr. Gibson graduated from Mississippi State University with a Ph.D. in physiology and holds a master's degree in zoology and a B.S. in general science from that same institution. He has served on the standing CMA Committee and its predecessor, the Stockpile Committee, because of his more than 25 years of experience in toxicology and occupational safety and health. He was appointed as the U.S. representative to the Scientific Advisory Board of the Organization for the Prohibition of Chemical Weapons in October 2003. He has served on a variety of chemical demilitarization ad hoc committees, including as chair of the Committee to Review and Assess Industrial Hygiene Standards and Practices at Tooele Chemical Agent Disposal Facility (TOCDF). He is currently the chair of the standing Committee on Chemical Demilitarization of the Board on Army Science and Technology.

Ronald M. Bishop is founder and president of AEHS, Inc., an environmental, health, and safety consulting services and training firm. He earned a B.S. from the University of Washington in preventive medicine (environmental health engineering) and a Master of Public Health from the University of Minnesota in industrial hygiene, with an additional concentration in air pollution. Mr. Bishop also served for 2 years as director of the Office of Safety and Health Protection at the Oak Ridge National Laboratory, where he was responsible for all safety, industrial hygiene, OSHA, hazardous waste management, and technical training. Mr. Bishop spent 25 years in the U.S. Army and held numerous positions in the environmental, safety, and health field, retiring as a colonel; the last assignment was as Commander of the U.S. Army Environmental Hygiene Agency. He has worked 12 years as an environmental, safety, and industrial hygiene consultant. This includes not only consulting per se but also teaching courses on indoor air quality, asbestos and lead, as well as respiratory protection, LO/TO, HAZCOM, confined space, and OSHA's 501 Voluntary Compliance. He served on the NRC Committee to Review and Assess Industrial Hygiene Standards and Practices at Tooele Chemical Agent Disposal Facility (TOCDF).

Colin G. Drury is Distinguished Professor of Industrial Engineering at the University at Buffalo, State University of New York, where his work concentrates on the application of ergonomics to manufacturing and maintenance processes. Formerly manager of ergonomics at Pilkington Glass, he has over 200 publications on topics in industrial process control, quality control, and aviation maintenance and safety. As the founding executive director of the Center for Industrial Effectiveness, he has worked with regional industries to improve competitiveness and has been credited with creating and saving thousands of jobs in the Western New York region. He is founding director of the Research Institute for Safety and Security in Transportation, applying human factors to error reduction in aviation security and inspection. Dr. Drury is a fellow of the Human Factors and Ergonomics Society, the Institute of Industrial Engineers, and the Ergonomics Society. He is a recipient of many awards, including the Bartlett Medal of the Ergonomics Society for his work in industrial quality control, the Human Factors and Ergonomics Society's Lauer award for safety, and the FAA's Excellence in Aviation Research Award. In 2006, Dr. Drury received the Andrew Roe Award of the American Association of Engineering Societies.

James H. Johnson, Jr., is a professor of civil engineering and dean of the College of Engineering, Architecture and Computer Sciences at Howard University. Prior to this appointment, he was the chair of the Department of Civil Engineering and interim associate vice president for research at Howard University. Dr. Johnson received a B.S. from Howard University, an M.S. from the University of Illinois, and a Ph.D. from the University of Delaware. He has taught undergraduate and graduate courses in the area of environmental engineering. Dr. Johnson's research interests include the treatment and disposal of hazardous substances, the evaluation of environmental policy issues in relation to minorities, the development of environmental curricula and strategies to increase the pool of underrepresented groups in the science, technology, engineering, and math disciplines. He is the past chair of the Board of Scientific Counselor's Executive Committee of the Environmental Protection Agency (EPA), a member of EPA's Science Advisory Board, and the co-principal investigator of the Department of Energy-sponsored HBCU/MI Environmental Technology Consortium. From 1989 to 2002, he was the associate director of the EPA-sponsored Great Lakes and Mid-Atlantic Center for Hazardous Substance Research; from 2005 to 2007, he served as a consultant to the Office of the President, University of California, as a member of the Environmental, Health and Safety Panel monitoring activities at the three DOE national laboratories operated by the university. Dr. Johnson is a member of the National Research Council's (NRC's) Division of Earth and Life Sciences Oversight Committee and chair of the Anne Arundel Community College board of trustees. Other recent service activities include membership of NRC's Board on Radioactive Waste Management and its Board on Environmental Studies and Toxicology, the board of directors of the Civil Engineering Research Foundation (CERF), and the Space Day Foundation. He also serves on several university and center advisory committees. Dr. Johnson is a registered professional engineer in the District of Columbia and a diplomat of the American Academy of Environmental Engineers. He is the 2005 recipient of the National Society of Black Engineers Lifetime Achievement Award in Academia and the 2008 Water Environmental Federation Gordon Maskew Fair Medal for significant contributions to the education and development of future engineers.

Randal J. Keller is currently a professor in the Department of Occupational Safety and Health of Murray State University. He received a B.A. in chemistry from Eisenhower College in 1979, an M.S. in toxicology from Utah State University in 1984, and a Ph.D., also in toxicology, from Utah State University in 1988. He is certified in the Comprehensive Practice of Industrial Hygiene by the American Board of Industrial Hygiene, the Comprehensive Practice of Safety by the Board of Certified Safety Professionals, and in the General Practice of Toxicology by the American Board of Toxicology. Dr. Keller is widely published and maintains an independent consulting practice related to toxicology, industrial hygiene, and safety. He served on the NRC's Committee to Review and Assess Industrial Hygiene Standards and Practices at the Tooele Chemical Agent Disposal Facility (TOCDF).

W. Monroe Keyserling, professor, University of Michigan, has 29 years' experience in research and teaching activities focused on occupational safety and health. He has taught courses in safety engineering methods, work measurement, prevention of work-related musculoskeletal disorders, and a seminar in occupational health and safety engineering. Dr. Keyserling holds a B.I.E. in industrial and systems engineering from the Georgia Institute of Technology and an M.S.E. in industrial and operations engineering, an M.S. in industrial health science, and a Ph.D. in industrial and operations engineering and industrial health from the University of Michigan. Dr. Keyserling has authored over 120 journal articles, book chapters, and technical reports. His primary research area has been developing methods and tools for measuring workplace exposures that increase the risk of work-related musculoskeletal disorders such as low back pain and carpal tunnel syndrome. Dr. Keyserling has also been active in university-wide and national efforts to promote multidisciplinary education and research in occupational health and safety. From 1995 to 2000, he served as director of the University of Michigan's Center for Occupational Health and Safety Engineering, a collaboration involving the College of Engineering, the School of Public Health, and the School of Nursing. From 1999 to 2002, he served as president of the Association of University Programs in Occupational Health and Safety (AUPOHS), a national advocacy group that promotes safety and health education and research.

Otis A. Shelton is associate director for the safety and environmental services compliance and operational assessments program for Praxair, Inc., a position he has held since 1992. In this position, Mr. Shelton is responsible for managing Praxair's assessment program, which focuses on the environmental, operational safety, personnel safety, industrial hygiene, emergency planning, distribution, and medical gases programs. Previously, Mr. Shelton managed Union Carbide Corporation's Regional Corporate Health, Safety, and Environmental Protection Audit Program, which reviewed UCC's health, safety, and environmental compliance in all of the company's operations worldwide. He holds an M.S. in chemical engineering from the University of Houston. He is a fellow of and has served on the board of directors of the American Institute of Chemical Engineers (AIChE) and served on the National Society of Black Engineers National Advisory Board for 20 years. He was elected Secretary of the AIChE in 2004.

Levi T. Thompson is Richard E. Balzhiser Professor of Chemical Engineering, professor of mechanical engineering, and director of the Hydrogen Energy Technology Laboratory at the University of Michigan. He earned a B.Ch.E. from the University of Delaware, M.S.E. degrees in chemical engineering and nuclear engineering, and a Ph.D. in chemical engineering from the University of Michigan. Research in his group focuses primarily on the design, characterization, and development of nanostructured catalytic, electrocatalytic, and adsorbent materials. In addition, his group is using micromachining and self-assembly methods to fabricate fuel processors, fuel cells, and batteries. From 2001 to 2005, he served as associate dean for undergraduate education in the College of Engineering and presently is director of the Michigan-Louis Stokes Alliance for Minority Participation. Professor Thompson was the recipient of a 2006 Michiganian of the Year Award for his research, entrepreneurship, and recruitment and mentoring of minority students, a National Science Foundation Presidential Young Investigator Award, the Engineering Society of Detroit Gold Award, the Union Carbide Innovation Recognition Award, and the Dow Chemical Good Teaching Award. He is also cofounder, with his wife, of T/J Technologies, a developer of nanomaterials for advanced batteries and a subsidiary of A123Systems. Professor Thompson is consulting editor for the *AIChE Journal* and a member of the External Advisory Committee for the Center of Advanced Materials for Purification of Water with Systems (an NSF Science and Technology Center at the University of Illinois), the National Academies' Chemical Sciences Roundtable, and the AIChE Chemical Engineering Technology Operating Council.

Lawrence J. Washington, the recently retired corporate vice president for Sustainability and Environmental Health and Safety (EH&S), worked for the Dow Chemical Company for over 37 years. Among his many distinctions, Mr. Washington chaired the Corporate Environmental Advisory Council, the EH&S Management Board, and the Crisis Management Team. He also served as an officer of the company. In previous roles, Mr. Washington served as corporate vice president, EH&S, Human Resources and Public Affairs. His career included many roles in operations including as leader of Dow's Western Division and general manager and site leader for Michigan operations. Mr. Washington earned bachelor's and master's degrees in chemical engineering from the University of Detroit.